ABYSS UNVEILED

*Secrets, Mysteries, and Wonders
of the Ocean Depths*

Copyright © 2024 O.Naumchyk
All rights reserved. No part of this
book may be reproduced, distributed, or transmitted in any
form or by any means, including photocopying, recording, or
other electronic or mechanical methods, without the prior
written permission of the author, except as allowed by
copyright law.
Every image, design, and piece of content within this book is
safeguarded by copyright and is the sole property of the
author.

CONTENTS OF THE BOOK:

Introduction
1-4

Fascinating, frightening, and amazing secrets the ocean holds
2-87

Final Chapter: The Eternal Abyss
88-89

INTRODUCTION

The ocean, vast and enigmatic, has always captured the human imagination. It is a world within a world, stretching far beyond the horizon, deeper than our most ambitious dreams, and teeming with life forms so alien they seem plucked from the pages of fantasy. This colossal, liquid expanse covers over 70% of the Earth's surface and yet, we have explored only a fraction of its depths. The ocean holds secrets that defy comprehension and mysteries that challenge the limits of human curiosity. It is a realm of contrasts—both serene and ferocious, nurturing and unforgiving, inviting and utterly terrifying.

For centuries, humanity stood at the edge of the shore, awestruck by the tides and waves, wondering what lay beyond. Early explorers ventured forth on fragile ships, driven by both fear and wonder. They returned with tales of uncharted lands and monstrous sea creatures—stories that blurred the line between reality and myth. Even now, in an age of satellites and submarines, much of the ocean remains uncharted. Its dark waters conceal more than just geographical mysteries; they guard the secrets of history, biology, and geology. From sunken ships and forgotten civilizations to the unrelenting power of underwater volcanoes and tectonic shifts, the ocean is a chronicle of Earth's past and a harbinger of its future.

The study of the ocean—oceanography—is a pursuit as endless as the sea itself. Scientists dive into its depths to unlock the riddles of its currents and ecosystems. Each expedition brings discoveries more astonishing than the last: hydrothermal vents spewing mineral-rich plumes into the abyss, trenches deeper than Mount Everest is tall, and vibrant coral reefs that host a kaleidoscope of life. There are creatures that glow in the dark, their bodies engineered by nature to thrive in crushing pressures and freezing temperatures. Fish that wield bioluminescent lures, jellyfish that may hold the secret to eternal life, and colossal squid with eyes the size of dinner plates all paint a picture of a world both beautiful and bizarre.

But the ocean is more than just a scientific frontier—it is a place of stories and legends. Tales of the kraken, leviathans, and phantom ships like the Flying Dutchman hint at our enduring fascination with the unknown. Even today, sonar captures eerie sounds in the deep, nicknamed "The Bloop" or "Julia," that defy easy explanation. Are they geological phenomena or evidence of life as yet undiscovered? The ocean guards its answers well.

Perhaps most haunting are the things the ocean has claimed. Countless ships and planes, lost in its vastness, rest silently beneath the waves. The Titanic is among the most famous, but it is far from alone. The Bermuda Triangle remains an enigma, where vessels vanish without a trace, leaving behind questions that remain unanswered. And yet, even in death, the ocean transforms—wreckage becomes a haven for marine life, turning tragedy into renewal.

The ocean also holds keys to understanding climate and survival. It regulates our planet's temperature, stores carbon, and provides half the oxygen we breathe. Its tides are governed by the moon, its currents by the rotation of the Earth, and its storms by the interplay of wind and water. Yet, this lifeblood of the planet is under threat, its delicate balance disrupted by human activity. Plastics and pollution seep into its depths, disrupting ecosystems that have thrived for millennia. The battle to protect the ocean is as crucial as the quest to explore it.

To step into the ocean is to step into a world of contrasts: beauty and danger, tranquility and chaos, knowledge and mystery. Each drop of water holds a story, each wave a secret, each unexplored trench the potential for discoveries that could redefine how we see our world. The ocean is not merely a part of our planet; it is its heart, its soul, and its greatest enigma. In this book, we will dive headlong into the wonders, horrors, and miracles of the deep, unraveling the stories that make the ocean one of the most fascinating frontiers in existence. Prepare to be amazed, unsettled, and captivated, for this is a journey into the unknown—a voyage into the incredible secrets of the ocean.

FASCINATING, FRIGHTENING, AND AMAZING SECRETS THE OCEAN HOLDS

1. The Bloop: A Mystery of the Deep

In 1997, an ultra-low frequency sound, now known as "The Bloop," was detected by NOAA's underwater listening stations. This mysterious noise was louder than any known biological source and seemed to emanate from the South Pacific, near an empty stretch of ocean. Speculations ran wild: Was it a colossal creature, far larger than any whale? A remnant of prehistoric life? Eventually, scientists proposed that the sound might have been caused by the fracturing of icebergs, but the exact origin remains unconfirmed. The mystery lingers, with some still holding out hope that the sound is evidence of something alive and unknown lurking in the abyss.

2. The Immortal Jellyfish

Imagine a creature that cheats death. The "immortal jellyfish" (Turritopsis dohrnii), a tiny, translucent being no larger than a fingernail, can revert to its polyp stage after reaching maturity, effectively restarting its life cycle. This process, akin to a butterfly turning back into a caterpillar, has been observed under laboratory conditions and may occur in the wild. While predators and disease can still kill these jellyfish, their cellular rejuvenation could hold the key to understanding aging in humans. Scientists continue to study its unique biology, hoping to unlock the secrets of immortality.

3. Mariana Trench: The Challenger Deep

The Mariana Trench, located in the western Pacific, is the deepest part of the world's oceans, plunging nearly 36,000 feet into the abyss. In 2012, filmmaker James Cameron became one of the few people to descend solo into the Challenger Deep, the trench's deepest point. What's most astonishing about this dark, crushing environment is that it teems with life. Bizarre creatures like amphipods the size of rats and translucent snailfish thrive there. These organisms have adapted to the extreme pressure, which would crush a human instantly. Scientists are still discovering new species in this enigmatic trench, hinting at how life might survive on other planets.

4. Underwater Volcanoes and the Birth of New Islands

Beneath the ocean's surface, volcanoes erupt with fiery violence, shaping the Earth's crust and sometimes birthing new islands. One of the most dramatic examples occurred in 2014, when an underwater eruption in the Pacific created Hunga Tonga-Hunga Ha'apai, a new island near Tonga. What's remarkable is that such events are relatively common; there are over a million underwater volcanoes, many still active. Some, like the Pacific Ring of Fire, pose serious threats to coastal populations. Studying these volcanoes helps scientists understand Earth's geology and the dynamic processes that shape our planet.

5. The Milky Sea Phenomenon

Sailors for centuries have reported encounters with "milky seas"—vast stretches of ocean glowing eerily under the cover of darkness. These phenomena, spanning hundreds of square miles, are caused by bioluminescent bacteria, specifically Vibrio harveyi. The bacteria communicate via quorum sensing, coordinating their light production to create a luminous spectacle. While scientists have managed to capture satellite images of these glowing waters, much about their triggers remains unknown. Milky seas are rare, but those lucky (or unlucky) enough to witness them describe a haunting, otherworldly experience.

6. The Lost City of Atlantis

The legend of Atlantis, a fabled city said to have sunk beneath the waves, has captivated imaginations for millennia. First described by the ancient Greek philosopher Plato, Atlantis was said to be an advanced civilization destroyed by divine punishment. While historians largely regard the tale as allegory, some researchers believe it could be rooted in real events, such as the volcanic eruption that destroyed the Minoan civilization on the island of Santorini. Despite numerous expeditions and hypotheses, Atlantis remains one of the ocean's most enduring mysteries, inspiring endless speculation about what might lie beneath the waves.

7. The Great Pacific Garbage Patch

Far from a myth, the Great Pacific Garbage Patch is a stark reminder of humanity's impact on the oceans. Spanning an estimated 1.6 million square kilometers, this swirling vortex of plastic debris lies between California and Hawaii. Much of the waste is microplastic, tiny fragments smaller than a grain of rice, which are ingested by marine life, entering the food chain. Efforts to clean up the patch, such as the Ocean Cleanup Project, are ongoing, but the challenge is enormous. Scientists warn that if left unchecked, plastic pollution could irreparably harm marine ecosystems.

8. Colossal Squid: The Real-Life Sea Monster

The colossal squid (Mesonychoteuthis hamiltoni) is one of the ocean's most elusive creatures, living at depths of up to 7,000 feet. Larger and more robust than the giant squid, it can grow up to 46 feet long and has the largest eyes of any known animal, up to a foot in diameter. Found primarily in Antarctic waters, this squid has razor-sharp hooks on its tentacles, adding to its fearsome reputation. Sightings are rare, but specimens retrieved from fishing nets have provided scientists with glimpses of its biology. Its behavior, including how it uses those enormous eyes in near-total darkness, remains largely unknown.

9. The Bermuda Triangle

This infamous stretch of the Atlantic Ocean, bounded by Miami, Bermuda, and Puerto Rico, has been the site of numerous unexplained disappearances. Ships and planes have vanished without a trace, fueling speculation about supernatural forces, extraterrestrial activity, or even time warps. While many of these incidents have plausible explanations, such as sudden storms or navigational errors, some cases, like the disappearance of Flight 19 in 1945, remain shrouded in mystery. Scientists attribute the phenomena to natural occurrences, such as methane hydrates destabilizing and causing ships to sink, but the allure of the Bermuda Triangle endures.

10. Zombie Worms: Nature's Deep-Sea Recyclers

In the crushing depths of the ocean, the bizarre "zombie worm" (Osedax) feasts on the bones of dead whales. These worms lack mouths or stomachs; instead, they secrete acid to break down bone, extracting nutrients with the help of symbiotic bacteria. Discovered in 2002, zombie worms play a crucial role in the deep-sea ecosystem, recycling organic matter. Their peculiar feeding habits and adaptations have intrigued scientists, who marvel at the strange ways life evolves to survive in the ocean's most extreme environments.

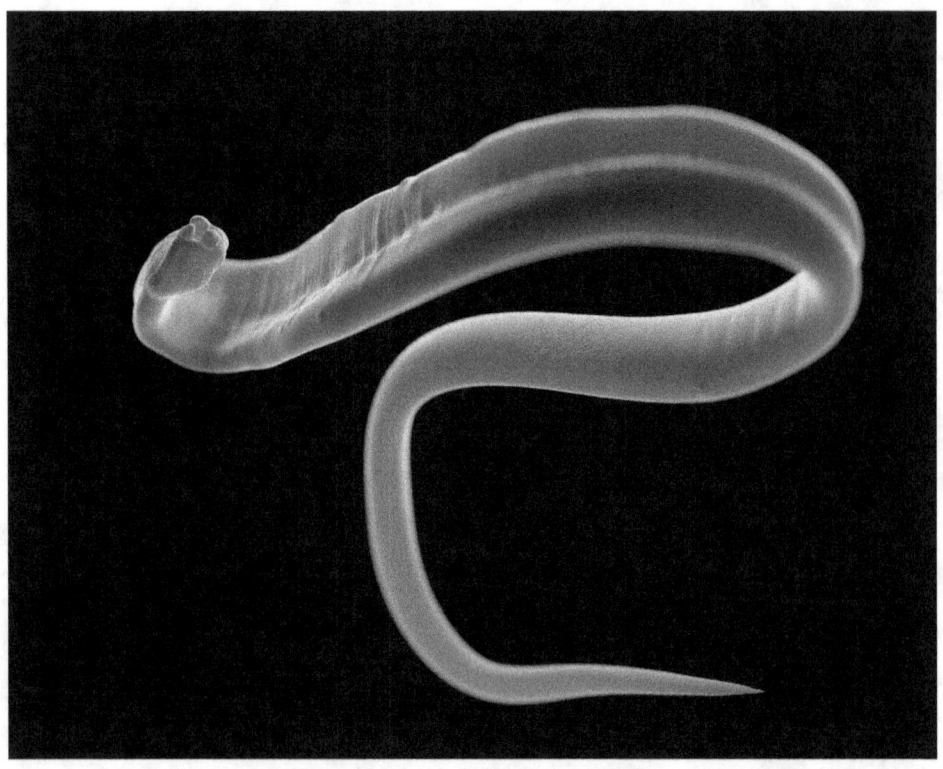

11. Dragonfish: The Nightmare Predator of the Deep

Lurking in the ocean's twilight zone, the dragonfish is a creature straight out of a nightmare. This small, eel-like predator uses its bioluminescent barbel—a glowing lure that dangles in front of its mouth—to entice prey in the crushing darkness. But what truly sets the dragonfish apart is its transparency and near-invisibility in the water. Unlike most deep-sea creatures, it has specialized black pigmentation in its teeth and skin that absorbs nearly all light, rendering it practically undetectable. Scientists studying its remarkable adaptations believe these traits could inspire advances in stealth technology.

12. The Yonaguni Monument: Nature or Lost Civilization?

Off the coast of Yonaguni, Japan, lies an underwater structure that has baffled scientists and archaeologists for decades. The Yonaguni Monument is a series of massive, geometric formations resembling pyramids, staircases, and terraces, submerged in 100 feet of water. Some believe it to be the remains of an ancient civilization, potentially over 10,000 years old, predating known history. Others argue that the structure is the result of natural geological processes, such as erosion and tectonic activity. Divers exploring Yonaguni often report an eerie feeling, as though the ruins whisper secrets of a forgotten age.

13. The Gulf Stream and the Power of Ocean Currents

The Gulf Stream, a powerful ocean current, has a profound impact on the climate and ecosystems of the Atlantic Ocean. It acts as a conveyor belt, transporting warm water from the tropics to the North Atlantic and returning cold water southward. This dynamic system influences weather patterns, making Europe far warmer than its latitude would suggest. However, scientists have discovered troubling evidence that the Gulf Stream is weakening, potentially due to climate change. A collapse of this current would have catastrophic consequences, plunging Europe into icy winters and disrupting marine life along its path. The ocean, it seems, wields power far beyond its waves.

14. Underwater Lakes and Rivers

Deep beneath the ocean's surface, in places like the Gulf of Mexico, lie underwater lakes and rivers—strange, alien landscapes where salt levels are so high that the water remains separate from the surrounding ocean. Known as brine pools, these bizarre phenomena are deadly to most marine life. Creatures that wander into them often succumb to the toxic, oxygen-deprived environment, creating eerie "graveyards" of crabs and fish. Yet, some hardy organisms, like tube worms and shrimp, thrive in these extreme conditions, leading scientists to speculate on life's potential in similarly harsh environments on other planets.

15. The Ghost Ships of the Ocean

The open sea has always been a place of mystery, and few tales are as chilling as those of ghost ships—vessels found adrift, abandoned by their crews, with no clear explanation. One of the most infamous cases is the Mary Celeste, discovered in 1872 near the Azores. Its crew was missing, but the ship was intact, with cargo and personal belongings undisturbed. Another is the SS Ourang Medan, a Dutch vessel reportedly found with its crew dead, expressions of terror frozen on their faces, before the ship mysteriously exploded and sank. While many ghost ship stories can be attributed to piracy, storms, or mutiny, others remain unsolved, feeding the enduring fear of the ocean's unknown forces.

16. The Abyssal Gigantism Phenomenon

In the deep ocean, where sunlight cannot penetrate, life takes on strange proportions. Scientists have discovered a phenomenon called abyssal gigantism, where deep-sea creatures grow to enormous sizes. Examples include giant isopods—crustaceans that resemble oversized pill bugs—and the colossal squid. This strange adaptation is thought to result from cold temperatures, high pressure, and limited food, which favor slower metabolisms and longer lifespans. Yet, the exact mechanisms remain a mystery. Abyssal gigantism challenges our understanding of evolution and hints at the vast, untapped potential of life in extreme environments.

17. The Mysterious Submarine Circles of Japan

Off the coast of Amami-Oshima, Japan, divers discovered intricate, circular patterns etched into the sandy seafloor. At first, the purpose and creator of these formations baffled researchers. It wasn't until 2011 that the mystery was solved: these "crop circles" were made by male pufferfish as part of an elaborate mating ritual. Using their fins, the fish sculpt symmetrical ridges and grooves, decorating the structure with shells to attract a mate. The discovery amazed scientists, showcasing the creativity and complexity of marine life in the most unexpected ways.

18. Bioluminescent Bays: Nature's Living Light Show

In Puerto Rico and other tropical locations, there are bays where the water lights up with a ghostly blue glow when disturbed. These bioluminescent bays are filled with Pyrodinium bahamense, microscopic plankton that emit light as a defense mechanism. Mosquito Bay, on the island of Vieques, is among the most famous, offering visitors a magical experience. However, these ecosystems are delicate, and changes in water quality or temperature can disrupt the plankton populations. Scientists continue to study these bays to understand how bioluminescence evolved and how to protect this dazzling natural wonder.

19. The Baltic Sea Anomaly

In 2011, a team of Swedish treasure hunters using sonar equipment discovered a strange object resting on the floor of the Baltic Sea. Shaped like a massive disc, the so-called Baltic Sea Anomaly sparked speculation about its origins. Some compared it to the Millennium Falcon from Star Wars, suggesting it might be an ancient UFO. Others believe it could be a natural geological formation or even a glacial deposit. Subsequent dives have yielded inconclusive results, leaving the anomaly shrouded in mystery. Whether natural or man-made, it serves as a reminder of how little we know about what lies beneath the waves.

20. The Case of the Disappearing Island

In 2012, an island in the South Pacific, known as Sandy Island, appeared on maps and was even documented in scientific databases. However, when researchers set out to study it, they found nothing but open ocean where the landmass was supposed to be. Dubbed the "phantom island," Sandy Island was eventually removed from maps, but its brief existence raises questions about human error, shifting tectonic plates, and the challenges of mapping the vast, ever-changing ocean.

21. The Underwater Waterfall of Mauritius

Off the coast of Mauritius, there exists a natural phenomenon so stunning that it defies belief: an underwater waterfall. When seen from above, the ocean appears to cascade downward into a bottomless void. This illusion is caused by sand and silt being carried by powerful underwater currents over the edge of a submerged shelf. While it's not a true waterfall, the effect is breathtaking. Satellite images and aerial photos capture this mesmerizing sight, but it's only visible from above—reminding us that the ocean hides its wonders until viewed from just the right perspective.

22. Lake Vostok: A Hidden World Beneath Antarctic Ice

Lake Vostok, one of the world's largest subglacial lakes, lies buried beneath 13,000 feet of Antarctic ice. This ancient body of water has been sealed off for millions of years, making it a time capsule of Earth's past. In 2012, Russian scientists drilled into the lake, collecting water samples that revealed microbial life surviving in complete isolation. These organisms thrive without sunlight, feeding off minerals in the ice. The discovery fuels speculation about life on icy moons like Europa, where similar conditions might exist. Studying Lake Vostok is like peering into another planet's ecosystem.

23. The Wreck of the Titanic: Time Capsule of Tragedy

More than a century after it sank in 1912, the wreck of the Titanic still haunts the ocean floor, nearly 12,500 feet below the surface. Discovered in 1985 by Dr. Robert Ballard, the site revealed a hauntingly preserved relic of history. Over the decades, scientists have monitored the wreck's slow decay, caused by metal-eating bacteria named Halomonas titanicae. While much of the ship remains intact, it is gradually being reclaimed by the ocean. Exploring the Titanic offers a glimpse into a tragedy frozen in time, a poignant reminder of human ambition and nature's immense power.

24. The Great Blue Hole: A Portal to the Past

Off the coast of Belize lies the Great Blue Hole, a circular sinkhole over 1,000 feet wide and 400 feet deep. This geological wonder is a diver's paradise, revealing stalactites and stalagmites formed when the area was a dry cave during the last Ice Age. Descending into its depths feels like traveling back in time, offering clues about ancient sea levels and climate. Jacques Cousteau famously called it one of the best dive sites in the world. However, its dark interior harbors dangers, and several divers have lost their lives exploring its enigmatic depths.

25. The "White Shark Café": A Shark Gathering Mystery

Every year, great white sharks migrate to a remote, mid-Pacific region known as the "White Shark Café." Located halfway between California and Hawaii, this area is puzzling because it offers no obvious food sources. Scientists have tagged and tracked sharks to uncover their behavior, finding that they dive to extreme depths, sometimes up to 3,000 feet, before returning to the surface. Why do they gather there?

Some researchers speculate it could be a mating ground or a stopover for feeding on deep-sea squid. The truth remains elusive, highlighting how much we have yet to learn about even the ocean's most iconic predators.

26. The Shadowy Realm of the Midnight Zone

Below 3,300 feet lies the ocean's midnight zone, a realm of perpetual darkness where sunlight cannot reach. Here, life exists under crushing pressure in water temperatures near freezing. Many creatures in this zone rely on bioluminescence to hunt, communicate, or evade predators. Among them is the anglerfish, which dangles a glowing lure to attract prey, and the barreleye fish, whose transparent head allows it to see upward to spot silhouettes of prey against faint light. Exploring the midnight zone has revealed a bizarre, alien world, pushing the boundaries of biology and resilience.

27. The Unsinkable Legend of the Mary Rose

The Mary Rose, a Tudor warship sunk in 1545, lay buried in the Solent for over 400 years until it was rediscovered in 1971. This historic ship offered an incredible snapshot of 16th-century life, preserving weapons, tools, and personal belongings in near-pristine condition due to the oxygen-poor seabed. Archaeologists were astonished by the intact remains of a dog thought to have been the ship's ratter. The Mary Rose is now displayed in Portsmouth, but her recovery was a race against time, as marine bacteria threatened to consume her wooden timbers. This wreck tells a tale of warfare, tragedy, and preservation.

28. Hydrothermal Vents: Chimneys of Creation

Discovered in 1977, hydrothermal vents are cracks in the seafloor that release superheated, mineral-rich water, creating towering structures known as "black smokers." Despite the extreme heat and toxicity, these vents support thriving ecosystems, including giant tube worms, clams, and crabs. The discovery of life in these harsh conditions shattered long-held assumptions that sunlight was essential for survival. Instead, organisms rely on chemosynthesis, harnessing energy from chemicals in the vent fluids. Hydrothermal vents provide insights into how life might exist on other planets, such as Jupiter's moon Europa or Saturn's moon Enceladus.

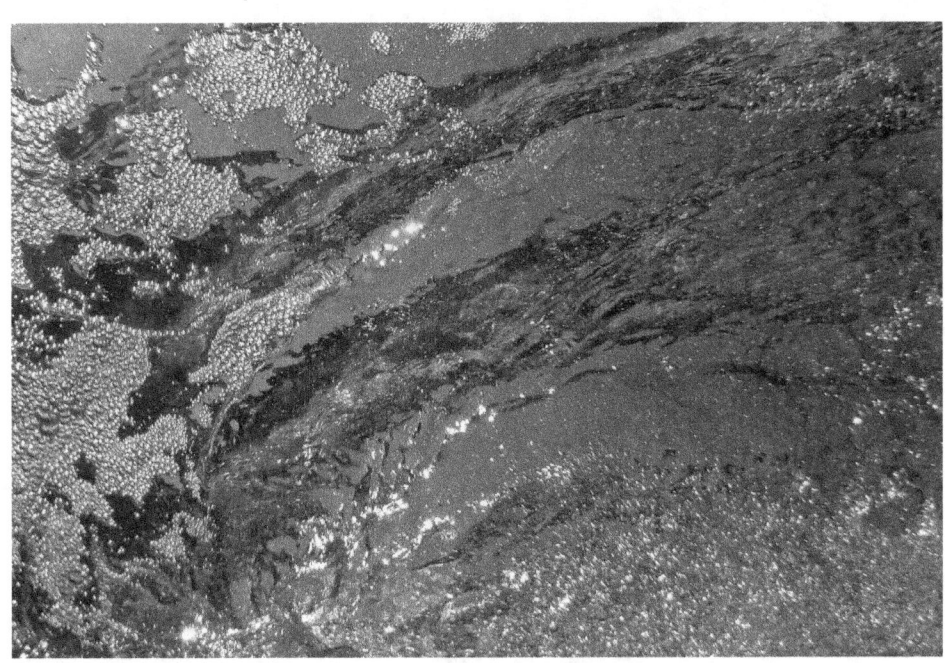

29. The Wreck of the USS Indianapolis

The USS Indianapolis met a tragic fate in July 1945, torpedoed by a Japanese submarine after delivering critical components of the atomic bomb. Of the 1,196 crew members aboard, nearly 900 survived the initial sinking, only to face days of exposure, dehydration, and shark attacks. The wreck was discovered in 2017, over 18,000 feet deep in the Pacific. Studying its remains has provided closure for families of the lost sailors and new insights into one of the U.S. Navy's worst disasters. The tale of the Indianapolis is both a story of heroism and a sobering reminder of the ocean's dangers.

30. The Devil's Sea: Japan's Bermuda Triangle

The Devil's Sea, or Dragon's Triangle, off Japan's coast, is steeped in legend. Similar to the Bermuda Triangle, this area has been associated with mysterious disappearances of ships and planes. Japanese folklore attributes these events to dragons or sea demons, while modern theories suggest undersea volcanic activity or rogue waves. In the 1950s, a research vessel, Kaiyo Maru No. 5, vanished in the area, further fueling the mystery. While many incidents have natural explanations, the eerie reputation of the Devil's Sea endures, keeping it one of the ocean's most ominous regions.

31. The Strangling Ghost Nets of the Ocean

"Ghost nets" are abandoned or lost fishing nets that drift through the ocean, continuing to trap marine life in a silent, deadly grip. These nets entangle everything from small fish to sea turtles, seals, and even whales, often causing a slow and agonizing death. Ghost nets can also accumulate other debris, forming massive floating hazards. One such net, nicknamed the "Great Pacific Ghost Net," measured over 80 feet long and weighed several tons when recovered. Scientists and conservationists are racing to address this issue, employing drones and specialized vessels to remove these deadly remnants from the water.

32. The Glowing Shark: A Predator that Lights the Way

The kitefin shark, discovered in 2021 off New Zealand's coast, has a remarkable feature: it glows in the dark. This deep-sea predator, growing up to six feet long, emits a blue-green bioluminescent light from its underbelly. Scientists believe this adaptation helps the shark camouflage itself against the faint light filtering from above, hiding its silhouette from prey below. It may also serve to lure smaller creatures or even signal to other sharks. The discovery of a glowing predator challenges our assumptions about the roles of bioluminescence in marine ecosystems, where light is often as crucial as food or oxygen.

33. Blackbeard's Sunken Treasure

The infamous pirate Blackbeard terrorized the seas in the early 18th century aboard his flagship, Queen Anne's Revenge. In 1718, the ship ran aground off the coast of North Carolina and was lost to history until its rediscovery in 1996. Since then, divers have recovered cannons, gold dust, medical tools, and even a pirate's cutlass from the wreck. These artifacts paint a vivid picture of pirate life, but the wreck has not yielded Blackbeard's fabled treasure—if it ever existed. The Queen Anne's Revenge remains a tantalizing link to a golden age of piracy, and some speculate its secrets have yet to be fully unearthed.

34. The Singing Sands of the Ocean Floor

Certain beaches around the world, like those in Hawaii and the Maldives, are known for their "singing sands," which produce musical tones when disturbed. But did you know the ocean floor can "sing" as well? Scientists have recorded strange, rhythmic vibrations emanating from the seabed in the Indian Ocean, likened to musical notes. While initially thought to be geological in origin, recent studies suggest these sounds might be caused by millions of small marine creatures moving or feeding in unison. These mysterious melodies hint at the hidden symphony of life thriving beneath the waves.

35. Megatsunamis: The Ocean's Ultimate Fury

While tsunamis caused by earthquakes are devastating, "megatsunamis" are on another scale entirely. These colossal waves, often over 1,000 feet tall, are triggered by sudden, massive geological events, such as underwater landslides or the collapse of volcanic islands. One of the most infamous examples occurred in 1958 in Lituya Bay, Alaska, when a landslide caused by an earthquake generated a wave over 1,700 feet high—the tallest in recorded history. Such events are rare but catastrophic, and scientists are monitoring regions like the Canary Islands, where volcanic instability could unleash a megatsunami with global consequences.

36. Vampire Squid: Master of the Deep

The vampire squid, despite its ominous name, is not a bloodthirsty predator. Instead, this deep-sea cephalopod survives by feeding on "marine snow," a mix of organic debris falling from the ocean's surface. Its red eyes, cloak-like webbing, and bioluminescent displays give it a haunting appearance. When threatened, it turns itself inside out, creating a glowing, spiky shield to confuse predators. Scientists studying the vampire squid marvel at its efficiency in surviving in one of the most nutrient-poor environments on Earth, making it a master of adaptation.

37. The Disappearance of Flight MH370

One of the most perplexing aviation mysteries of modern times is the disappearance of Malaysia Airlines Flight MH370 in 2014. The plane vanished over the Indian Ocean with 239 people aboard, sparking an international search. Debris confirmed to be from the aircraft washed up on islands thousands of miles apart, but the main wreckage remains undiscovered. Theories about the cause range from technical failure to pilot actions, but without the flight recorders, the mystery remains unsolved. The incident highlights the vastness of the ocean, where even something as large as a Boeing 777 can vanish without a trace.

38. Sea Serpents and the Real-Life Oarfish

For centuries, sailors have reported sightings of massive, snake-like sea creatures—"sea serpents" that slither through the waves. While many accounts are likely exaggerated, some may be inspired by encounters with the oarfish. These remarkable creatures can grow up to 36 feet long and are the longest bony fish in the world. Rarely seen alive, oarfish are deep-sea dwellers that occasionally surface when sick or dying, fueling legends of sea monsters. Their shimmering, ribbon-like bodies and undulating movements make them one of the most otherworldly creatures ever documented.

39. The Ocean's Hidden Gold Reserves

It's estimated that the ocean contains about 20 million tons of dissolved gold. Unlike gold deposits on land, this gold is spread out in microscopic particles, making it almost impossible to mine profitably with current technology. However, some marine mining companies are exploring ways to extract gold from seafloor sediments, where it collects around hydrothermal vents. The idea of "golden oceans" has inspired dreams of untold wealth, but it also raises concerns about the environmental impact of deep-sea mining, which could disrupt fragile ecosystems.

40. The Tragic Tale of the Essex: Inspiration for Moby-Dick

In 1820, the whaleship Essex was rammed and sunk by a sperm whale in the Pacific Ocean—a real-life event that inspired Herman Melville's classic novel, Moby-Dick. The surviving crew endured a harrowing ordeal, drifting for months in small boats. Starvation led some to resort to cannibalism, making their story one of the most chilling maritime survival accounts. The wreck of the Essex has never been found, but its legacy endures as a reminder of humanity's fragile existence against the ocean's immense power and the creatures that inhabit it.

41. Cleopatra's Sunken City: The Lost Treasures of Alexandria

In the shallow waters off the coast of Egypt lies a city lost to time: the ancient metropolis of Heracleion, often associated with Cleopatra's Alexandria. For centuries, this bustling hub of trade and culture lay submerged, hidden beneath layers of silt and sand. Discovered in 2000, the underwater city revealed grand statues, coins, jewelry, and the ruins of temples. Archaeologists believe the city sank due to earthquakes and rising sea levels in the 8th century AD. Exploring Heracleion is like stepping into a time capsule, offering a rare glimpse into ancient Egyptian life, trade, and religion.

42. Ocean Storms: The Fury of Super Typhoons

The ocean is the birthplace of Earth's most ferocious storms: typhoons, hurricanes, and cyclones. These immense systems can release as much energy in a single day as all the world's nuclear weapons combined. One of the most extreme examples is Typhoon Tip, which formed over the Pacific in 1979. Tip grew to a staggering 1,380 miles in diameter, with wind speeds of 190 mph. What makes these storms even more terrifying is how they intensify over warm ocean waters, driven by climate change. Modern satellites and drones monitor these storms, but their sheer destructive power continues to test human resilience.

43. The Coelacanth: A Living Fossil Resurfaces

Thought to have gone extinct 66 million years ago, the coelacanth was rediscovered alive and well in 1938 off the coast of South Africa. This prehistoric fish, with its lobed fins and ancient lineage, represents a living fossil, providing scientists with a rare window into evolutionary history. Coelacanths live in deep, dark underwater caves, often at depths of up to 2,300 feet. Their slow metabolism and unique ability to detect electrical fields help them navigate and survive in this extreme environment. The coelacanth reminds us that the ocean can still surprise us with creatures thought lost to time.

44. The Haunted Shipwreck of the SS Thistlegorm

Resting in the Red Sea lies the SS Thistlegorm, a British cargo ship sunk by German bombers during World War II. Discovered by Jacques Cousteau in the 1950s, this wreck is a haunting underwater museum, its hold filled with wartime supplies: trucks, motorcycles, rifles, and even boots. Divers exploring the Thistlegorm report an eerie sense of history frozen in time, as if the ship's crew might return at any moment. Coral has begun to reclaim the wreck, transforming it into a thriving reef teeming with fish and marine life, blurring the line between human history and nature's reclamation.

45. The Colossal Hurricane Katrina

When Hurricane Katrina struck in 2005, it brought more than just destruction to the Gulf Coast—it highlighted the power and unpredictability of ocean-driven weather. This Category 5 hurricane generated waves over 90 feet high and storm surges that overwhelmed New Orleans' levees, flooding 80% of the city. Katrina's impact extended beyond human lives, reshaping coastlines, destroying wetlands, and affecting marine ecosystems. The storm's ferocity was fueled by warm ocean waters, a reminder of how climate change intensifies extreme weather. Scientists study storms like Katrina to better predict and mitigate future disasters.

46. Zombie Crabs and the Parasite that Controls Them

Deep in the ocean, parasitic barnacles of the genus Sacculina have evolved a horrifying way to survive: by taking over the bodies of crabs. The parasite invades a crab's shell and spreads its tendrils into the host's nervous system, effectively controlling its behavior. The infected crab stops reproducing and instead cares for the parasite's eggs as if they were its own. This bizarre life cycle has fascinated scientists, who study Sacculina to understand how parasites manipulate their hosts. The zombie-like transformation of crabs shows that even in the ocean, nature can take a turn for the macabre.

47. The Ghost Fleet of Truk Lagoon

In 1944, a devastating U.S. air attack during World War II sank dozens of Japanese ships in Truk Lagoon, creating one of the largest underwater graveyards in the world. Today, this lagoon in Micronesia is a haunting site, with rusting ships, planes, and tanks scattered across the seafloor. Coral and marine life have transformed these wrecks into vibrant artificial reefs, but artifacts like sake bottles, helmets, and human remains make it clear these ships are tombs. Divers visiting Truk Lagoon describe it as both beautiful and eerie, a stark reminder of war's toll.

48. The Gigantic Sea Spider

Sea spiders, or pycnogonids, are bizarre arthropods found in the deep ocean. Unlike their land-dwelling counterparts, some sea spiders grow to monstrous sizes, with leg spans exceeding two feet. Their hollow bodies contain digestive systems that extend into their legs, and they lack lungs, relying on diffusion to absorb oxygen. Found near hydrothermal vents and in polar waters, these creatures exemplify the strange adaptations needed to survive in extreme marine environments. Despite their name, sea spiders are not true spiders, but their alien-like appearance makes them one of the ocean's most peculiar inhabitants.

49. The Underwater City of Pavlopetri

Pavlopetri, off the coast of Greece, is the world's oldest known underwater city, dating back over 5,000 years. Discovered in 1967, this submerged town features roads, houses, courtyards, and even a complex water management system. Archaeologists believe rising sea levels slowly submerged the city, preserving it in remarkable detail. Using advanced 3D mapping, researchers have reconstructed Pavlopetri virtually, revealing a sophisticated Bronze Age settlement. This ancient city challenges our understanding of early human engineering and trade, offering a glimpse into life before history books were written.

50. Dragonfly Squid: The Glittering Jewel of the Deep

The dragonfly squid, also known as the jewel squid, is a dazzling deep-sea creature adorned with light-producing photophores across its body. These bioluminescent displays serve as camouflage, lures for prey, or signals to mates. The squid's oversized, reflective eyes enhance its ability to detect faint light in the deep ocean, where it lives at depths of 2,000 to 3,000 feet. Observing the dragonfly squid is like encountering a living gemstone, a reminder of the ocean's ability to create beauty in the most inhospitable environments.

51. The Ocean Floor's Upside-Down Waterfalls

In the ocean's depths, near underwater volcanoes and hydrothermal vents, scientists have observed a phenomenon that defies logic: what appear to be "upside-down waterfalls." These occur when superheated water, rich with dissolved minerals, gushes out of the seafloor and rises into the colder, denser ocean water above. The hot water plumes, laden with shimmering particles, look like inverted waterfalls cascading upward. The minerals released by these vents eventually settle, forming towering chimney-like structures called black smokers. These surreal landscapes are some of the most otherworldly sights on Earth and may mirror conditions on alien worlds.

52. The Abyssal Plains: The Ocean's Deserts

At depths of 10,000 to 20,000 feet, the abyssal plains are some of the flattest and most monotonous regions on Earth. Despite their barren appearance, these vast stretches of ocean floor are critical to global ecosystems, serving as repositories for organic material that falls as "marine snow" from above. Surprisingly, life thrives here, with creatures like sea cucumbers, brittle stars, and xenophyophores—giant, single-celled organisms—scattered across the sediment. These plains also hide fossils and sediment records that tell the story of Earth's climate and geological history over millions of years.

53. The Lost Shipwrecks of the Deep Sea

The ocean floor is a vast graveyard of shipwrecks, many of which are yet to be discovered. Among the most remarkable finds is the wreck of the USS Johnston, a U.S. Navy destroyer sunk during World War II. In 2021, researchers found it at an astonishing depth of 21,180 feet in the Philippine Sea—the deepest shipwreck ever located. Despite its age, the ship's guns and hull were remarkably intact, offering a time capsule of history. Thousands of wrecks like the Johnston remain hidden in the deep, their stories waiting to resurface.

54. The Underwater Canyon of Monterey Bay

Monterey Bay, California, is home to an underwater canyon deeper than the Grand Canyon. Plunging over 10,000 feet below the surface, it is a hotspot of biodiversity, attracting species like giant squid, deep-sea octopuses, and even rare beaked whales. Scientists have deployed remote-operated vehicles (ROVs) to explore the canyon, discovering ghostly sponges and coral forests. The nutrient-rich waters of the canyon support a thriving ecosystem, demonstrating how even the deepest parts of the ocean floor can teem with life.

55. The Enigma of Missing Submarines

Numerous submarines lost during wartime still lie undiscovered on the ocean floor, their locations shrouded in mystery. One such case is the USS Scorpion, a nuclear submarine that vanished in the Atlantic in 1968 with 99 crew members aboard. Found five months later, its wreckage remains a subject of intrigue and speculation. Some theories suggest mechanical failure, while others hint at Cold War espionage. These deep-sea wrecks are not just military relics—they are tombs, preserved in the icy darkness, silent witnesses to history.

56. Deep-Sea Coral Forests: The Hidden Ecosystems

Far beneath the waves, at depths exceeding 3,000 feet, deep-sea coral forests thrive in the cold, dark waters. Unlike their shallow-water counterparts, these corals grow at a glacial pace, sometimes surviving for thousands of years. One of the largest known deep-sea coral reefs, off the coast of Norway, spans an area the size of Manhattan. These ecosystems provide habitat for countless species, many of which are yet to be identified. Studying these corals helps scientists understand how marine life adapts to extreme conditions and how climate change affects deep ocean biodiversity.

57. Methane Seeps: The Ocean Floor's Fire Ice

Methane seeps, or cold seeps, are places where methane gas escapes from the seafloor, often forming strange, icy structures known as methane hydrates. These deposits look like blocks of ice but can burn if ignited. Methane seeps are hotspots for unique life forms, including methane-eating bacteria and tube worms. Scientists are closely monitoring these seeps, as melting hydrates due to warming ocean temperatures could release vast amounts of methane—a potent greenhouse gas—into the atmosphere, potentially accelerating climate change.

58. The Secret Sounds of the Ocean Floor

Hydrophones placed on the seafloor have picked up mysterious, unexplainable sounds that spark both scientific curiosity and wonder. The "Upsweep" and "Slow Down" are examples of oceanic sounds recorded by NOAA's underwater listening network. These sounds are often associated with volcanic activity or shifting tectonic plates, but some remain unexplained. The hydrophones also capture the songs of whales, the clicking of dolphins, and the eerie grinding of icebergs scraping against the seafloor, creating a rich acoustic tapestry that reveals the hidden life of the ocean.

59. The South Sandwich Trench: A Hidden Depths Discovery

One of the lesser-known ocean trenches, the South Sandwich Trench, reaches depths of nearly 27,000 feet in the South Atlantic. In 2021, scientists exploring the trench using deep-sea landers discovered creatures adapted to extreme conditions, including amphipods and snailfish. The trench lies near volcanic islands, combining the crushing pressure of the deep with hydrothermal activity. This unique environment could hold clues to the origin of life on Earth, as it resembles the primordial conditions of early oceans billions of years ago.

60. The Titanic's Silent Companions

The wreck of the Titanic, lying at 12,500 feet on the North Atlantic seabed, is not alone. The surrounding area is littered with relics from the ship, including dishes, shoes, and even unopened champagne bottles. Intriguingly, the wreck has also become a habitat for strange deep-sea creatures, including crustaceans and microbial colonies that feed on the ship's iron. These bacteria, nicknamed "rusticles," are slowly consuming the Titanic, and scientists estimate the wreck may disappear entirely within the next 50 years.

61. The Dragon's Breath in the Arctic: Brinicles

Deep beneath the icy waters of the Arctic and Antarctic, a rare and deadly phenomenon occurs: the formation of brinicles, or "ice stalactites." These icy fingers grow downward from the surface as super-cold, salty water sinks and freezes everything in its path. Dubbed the "ice finger of death," a brinicle can trap and freeze sea creatures like starfish and sea urchins that unwittingly drift into its flow. First documented in 2011, brinicles are not only mesmerizing but also highlight the extremes of oceanic physics in polar environments.

62. The Orda Cave: An Underwater Labyrinth

In Russia lies the Orda Cave, one of the world's largest underwater gypsum caves, stretching over three miles. Its labyrinthine corridors and crystal-clear waters make it a magnet for cave divers, but its eerie beauty hides immense danger. The cave is pitch black and bone-chillingly cold, with narrow passageways that require expert navigation. Divers have reported hearing strange sounds and feeling disoriented, adding an air of mystery to the cave. The gypsum walls reflect light in dazzling patterns, making the Orda Cave feel like an underwater cathedral.

63. The Missing Oil of the Deepwater Horizon Spill

In 2010, the Deepwater Horizon oil spill released millions of barrels of oil into the Gulf of Mexico, creating one of the worst environmental disasters in history. While much of the oil was cleaned or evaporated, scientists discovered that vast amounts had sunk to the ocean floor, covering coral reefs and seafloor ecosystems in a toxic sludge. This hidden oil continues to impact marine life, from deep-sea crabs to microscopic plankton. The spill highlighted how events at the surface can have long-lasting consequences for the delicate ecosystems of the ocean floor.

64. The Yonaguni Underwater Pyramid

Off the coast of Yonaguni, Japan, lies a mysterious underwater structure resembling a pyramid. Discovered in 1987, the formation features sharp edges, steps, and terraces, leading some to speculate that it's the ruins of an ancient civilization. However, geologists argue that the structure may be a natural formation created by tectonic activity and erosion. Divers exploring Yonaguni report a sense of awe, as if the stones themselves whisper ancient secrets. Whether man-made or natural, the site continues to captivate archaeologists and adventurers alike.

65. The Lost Nuclear Submarine K-129

In 1968, the Soviet submarine K-129 sank mysteriously in the Pacific Ocean. The U.S. launched a secret mission, codenamed Project Azorian, to recover the wreck. Using a specially built ship, the Glomar Explorer, the mission succeeded in retrieving parts of the submarine from 16,000 feet below. Among the recovered items were Soviet nuclear missiles, fueling Cold War intrigue. The exact cause of K-129's sinking remains unknown, but its wreck has become a symbol of the ocean's role in espionage and geopolitical tension.

66. Alien Landscapes: Mud Volcanoes on the Ocean Floor

In certain regions of the ocean floor, like the Barbados Accretionary Prism, mud volcanoes bubble with methane and other gases, creating surreal underwater landscapes. These volcanoes don't spew lava but instead eject thick, oozing mud and hydrate crystals. They are often surrounded by unique life forms, such as clams and tubeworms, that rely on methane for sustenance. Studying mud volcanoes gives scientists a better understanding of geological processes beneath the seafloor, as well as insights into the potential for similar features on planets like Mars.

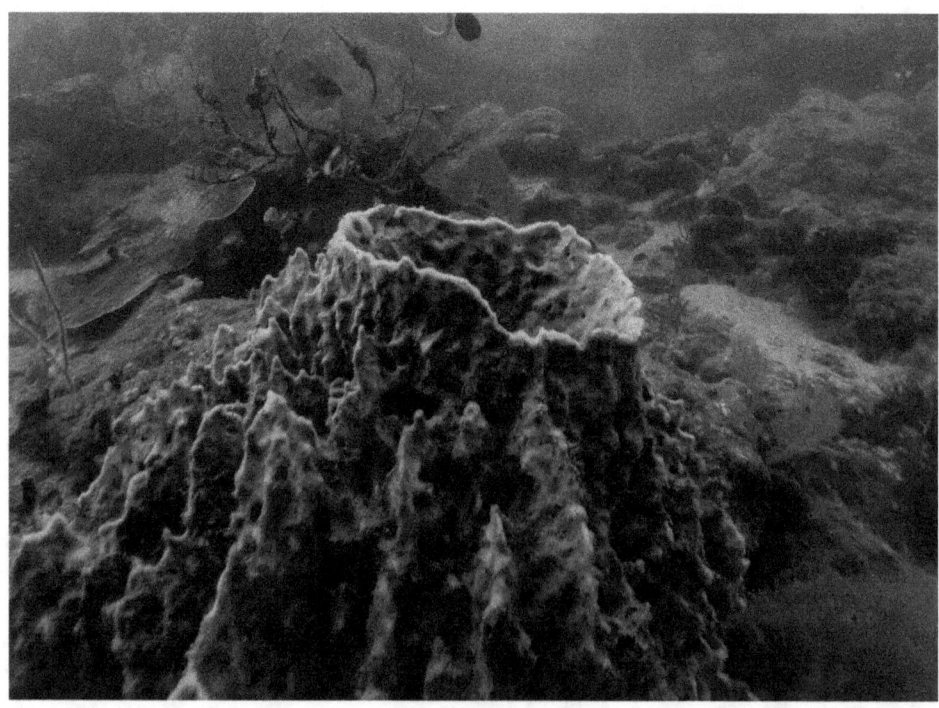

67. Zombie Worm Graveyards

Beneath the ocean floor, thousands of whale skeletons create unique "whale fall" ecosystems. These skeletons attract zombie worms (Osedax), which burrow into the bones and dissolve them using acid to feed on nutrients. These worms lack mouths and stomachs, relying on symbiotic bacteria to digest the marrow. Scientists studying whale falls have found that these bizarre ecosystems can last for decades, forming an important part of the deep-sea nutrient cycle. Each skeleton becomes a bustling metropolis for crabs, snails, and other scavengers in one of the ocean's strangest recycling systems.

68. The Underwater Lake in the Gulf of Mexico

In 2014, researchers discovered an underwater lake in the Gulf of Mexico nicknamed "Jacuzzi of Despair." This brine pool is five times saltier than the surrounding seawater and toxic to most marine life, forming a distinct boundary that looks like a shimmering lakebed. Fish or crabs that wander into the brine pool often don't survive. Despite its toxicity, the pool is home to extremophiles—microorganisms adapted to extreme conditions.

69. Deep-Sea Rift Zones: The Earth's Cracks Beneath the Sea

Along tectonic plate boundaries, such as the Mid-Atlantic Ridge, deep-sea rift zones mark where the Earth's crust is pulling apart. These zones are volcanic hotbeds, spewing lava that cools into vast basalt plains. Rift zones also house hydrothermal vents that teem with unique life forms, from giant tube worms to yet-undiscovered microorganisms. Studying these zones helps scientists understand plate tectonics and the origins of Earth's oceans.

70. The Deep Scattering Layer: The World's Largest Migration

Every night, billions of marine creatures, including fish, squid, and jellyfish, ascend from the ocean's depths to feed near the surface. This phenomenon, known as the deep scattering layer, is the largest migration on Earth, involving countless animals moving up and down thousands of feet. First discovered by sonar during World War II, this layer reflects sound waves, making it appear as if the ocean floor is moving. Scientists study this migration to understand its impact on the global carbon cycle.

71. The Midnight Paradox: Oceanic Dead Zones

Scattered across the ocean floor are "dead zones," areas where oxygen levels are so low that most marine life cannot survive. These regions are often caused by nutrient overloads from human activities, such as agricultural runoff, which lead to massive algal blooms. When the algae die and decompose, they consume oxygen, creating suffocating conditions. One of the largest dead zones is in the Gulf of Mexico, spanning up to 8,000 square miles. Yet, in these lifeless waters, some microorganisms thrive, offering clues about how life might persist in oxygen-poor environments elsewhere in the universe.

72. The Lost Fleet of Kublai Khan

In 1274 and 1281, Kublai Khan, the Mongol emperor, attempted to invade Japan with massive fleets of ships. Both times, his armadas were destroyed by powerful typhoons, which the Japanese called "kamikaze," or divine winds. For centuries, the wreckage of these fleets remained a mystery. In recent decades, archaeologists have uncovered pieces of Kublai Khan's ships off the coast of Japan, including anchors, weapons, and timbers.

73. The Dragon Hole: The Deepest Blue Hole

Located in the South China Sea, the Dragon Hole, or "Longdong," is the world's deepest blue hole, descending over 1,000 feet into the seafloor. Blue holes are underwater sinkholes, often surrounded by myths and legends. The Dragon Hole's waters are anoxic (oxygen-free) at depth, making it inhospitable to most life, but extremophiles have been found living there. Ancient fossils and coral deposits in the hole's walls are a geological time capsule, preserving clues about sea levels and climate changes over thousands of years.

74. Glass Sponges: The Ocean's Natural Engineers

At the bottom of the ocean, glass sponges form intricate, lattice-like skeletons made of silica, resembling delicate works of art. These sponges, some over 10,000 years old, are masters of water filtration, cycling thousands of liters of water daily. Found in deep-sea environments, they often form massive reefs, like those discovered off the coast of British Columbia. These reefs provide habitats for countless marine species and play a crucial role in maintaining the health of deep-sea ecosystems.

75. The Mariana Trench's Alien Slime Eels

The Mariana Trench, the deepest part of the ocean, is home to some of the most bizarre creatures on Earth, including hagfish, also known as slime eels. These ancient, jawless fish excrete massive amounts of slime when threatened, clogging the gills of predators and suffocating them. Living at depths of up to 8,000 feet, hagfish feed by burrowing into the carcasses of dead whales and fish, consuming them from the inside out. Their unique defense mechanism and scavenger lifestyle make them one of the ocean's most resilient inhabitants.

76. The Bermuda Triangle's Underwater Caves

Beneath the surface of the Bermuda Triangle lies an intricate network of underwater caves and sinkholes, known as blue holes. These geological formations are linked to the area's reputation for mysterious disappearances. Divers exploring blue holes in the Bahamas have discovered ancient stalactites, fossilized animal remains, and even human artifacts.

77. The Methane Bomb: Clathrates on the Seafloor

Buried in the sediments of the ocean floor are vast reserves of methane clathrates, crystalline structures that trap methane gas within water molecules. These "frozen time bombs" contain more carbon than all the world's fossil fuels combined. If released due to rising ocean temperatures, this methane could trigger catastrophic climate change. Scientists are racing to understand how stable these deposits are, and some even propose extracting methane as a new energy source. However, tampering with these volatile reserves could have dire consequences for the planet.

78. The Singing Whales of the Abyss

Blue whales, the largest animals on Earth, produce haunting songs that can travel thousands of miles through the ocean. These low-frequency sounds, sometimes below the range of human hearing, are thought to aid in communication, navigation, and mating. In the depths of the ocean, where light fades and pressure mounts, sound becomes the primary mode of interaction. Recent studies reveal that blue whale songs have changed over time, possibly due to shifts in ocean noise pollution. These songs remind us that even in the vast silence of the abyss, life finds a way to connect.

79. Deep-Sea Volcanoes and Pillow Lava

In the ocean's darkest depths, underwater volcanoes erupt with molten rock that cools instantly upon contact with water, forming unique structures called pillow lava. These bulbous formations create new seafloor, shaping the Earth's crust. One of the largest volcanic regions, the East Pacific Rise, is constantly expanding, adding several inches of new seafloor each year. These undersea eruptions also release minerals that feed deep-sea ecosystems, proving that even in fiery chaos, the ocean fosters life.

80. The Black Coral Forests of Hawaii

Deep in the waters surrounding Hawaii, forests of black coral grow on the ocean floor. Unlike their colorful shallow-water counterparts, black corals are dark and skeletal, with some specimens dating back over 4,000 years, making them among the oldest living organisms on Earth. These corals provide habitats for a variety of deep-sea species, from crabs to shrimp. However, black coral forests are slow-growing and vulnerable to deep-sea fishing and climate change. Protecting them is crucial to preserving the biodiversity of the deep ocean.

81. The Wreck of the Endurance: Shackleton's Lost Ship

In 1915, Ernest Shackleton's Antarctic expedition ship, the Endurance, was crushed by ice and sank beneath the Weddell Sea. For over a century, the ship's location remained a mystery due to the treacherous ice-covered waters. In 2022, explorers finally discovered the wreck, astonishingly well-preserved at a depth of 10,000 feet. The icy waters protected the wooden hull from decay, and the ship seemed frozen in time, with its name clearly visible on the stern. This discovery provided a tangible connection to one of the greatest survival stories in history and the indomitable human spirit.

82. The Ocean's Black Holes: Rogue Whirlpools

Much like their cosmic counterparts, the ocean has "black holes"—powerful whirlpools or eddies that trap water and debris within their boundaries. These phenomena, such as those in the Gulf Stream or near the Agulhas Current off South Africa, can persist for months and span hundreds of miles. Inside these whirlpools, marine life can become trapped, creating micro-ecosystems.

83. Deep-Sea Ghost Sharks: The Elusive Chimaeras

Ghost sharks, or chimaeras, are some of the ocean's most mysterious creatures, residing at depths of over 6,000 feet. These cartilaginous fish are relatives of sharks and rays, but they look otherworldly, with translucent bodies, glowing eyes, and long whip-like tails. Some species even have retractable sexual appendages on their foreheads, earning them the nickname "alien sharks." Scientists know little about their behavior due to their deep-sea habitat, but their adaptations showcase the incredible diversity of life in the darkest corners of the ocean.

84. Underwater Volcanoes That Create Acid Oceans

Near the Solomon Islands lies Kavachi, an active underwater volcano known as "Sharkcano" for its surprising inhabitants—sharks swimming in its acidic, ash-filled waters. The volcano regularly erupts, creating extreme conditions that would seem inhospitable. Yet, species like hammerhead and silky sharks thrive in the toxic environment, baffling scientists. Studying Kavachi and its inhabitants could reveal how marine life adapts to extreme changes, offering clues about survival in a warming world.

85. The Titanic's Sister Ship: The Britannic

Less famous than the Titanic, the Britannic was the third of the Olympic-class ships and served as a hospital ship during World War I. In 1916, it struck a German mine and sank in the Aegean Sea. The wreck was discovered in 1975 at a depth of 400 feet, its hull remarkably intact. Unlike the Titanic, most of the Britannic's crew survived, but its sinking highlighted the dangers of wartime naval operations. The Britannic remains one of the largest shipwrecks in the world and a haunting reminder of history's tragedies.

86. The Underwater Rivers of the Black Sea

Beneath the Black Sea's surface flows an underwater river, complete with channels, rapids, and waterfalls. This phenomenon occurs because of highly saline water flowing along the seafloor, creating a distinct, denser layer. The underwater river is so powerful that it has carved its own valleys and banks. If it were on land, this river would rank among the largest on Earth. Exploring these hidden waterways provides insights into ancient geological processes and the unique hydrodynamics of the ocean floor.

87. The Deep-Sea "Dumbo" Octopus

Named for its ear-like fins that resemble Disney's Dumbo, this deep-sea octopus lives at depths of up to 13,000 feet. These gentle creatures use their fins to gracefully "fly" through the water. Dumbo octopuses are adapted to the crushing pressures and cold temperatures of the abyss, feeding on crustaceans and worms found in the sediment. Unlike many octopus species, they lack an ink sac, as their deep-sea habitat offers little need for defense against predators. The Dumbo octopus exemplifies the quiet elegance of life in the deep.

88. Hydrothermal Vent Shrimp: Living in Boiling Water

In the extreme environment of hydrothermal vents, where temperatures can reach 750°F, lives the Alviniconcha shrimp. These tiny crustaceans cluster near vents, feeding on bacteria that thrive in the scalding waters. Their shells are coated in iron, giving them a metallic sheen, and they host symbiotic bacteria in their gills to process the chemicals spewed by the vents. These shrimp showcase the resilience of life, surviving where few other organisms can endure.

89. The Hidden Coral Reef Beneath the Amazon River

In 2016, researchers discovered a massive coral reef system at the mouth of the Amazon River, spanning nearly 600 miles. This reef, hidden beneath murky freshwater outflow, defied expectations, as corals typically thrive in clear, salty waters. The reef hosts sponges, fish, and unique coral species adapted to low light and high sedimentation. Scientists were shocked to find such biodiversity in an environment thought inhospitable for coral reefs, highlighting the ocean's capacity for unexpected wonders.

90. The Great South China Sea Ship Graveyard

The South China Sea is home to one of the largest concentrations of ancient shipwrecks in the world, with some dating back over a thousand years. These wrecks, many carrying cargos of ceramics, spices, and gold, were part of the ancient maritime Silk Road that connected Asia, the Middle East, and Europe. The most famous find, the Belitung Shipwreck, carried Tang Dynasty treasures and offers a snapshot of international trade during the 9th century.

91. The Colossal Sea Stars of the Deep

Giant sea stars, some over three feet across, roam the ocean floor at depths of up to 7,000 feet. These massive echinoderms are predators and scavengers, preying on clams, crabs, and even other sea stars. Their ability to regenerate lost arms makes them resilient hunters, while their vibrant red and orange hues create a striking contrast in the dim light of the deep ocean. Giant sea stars are vital to the ecosystem, maintaining balance by controlling populations of their prey.

92. The Gulper Eel: The Deep's Expanding Predator

In the ocean's midnight zone lives the gulper eel, a creature with an absurdly large mouth that can inflate like a balloon. This adaptation allows it to swallow prey much larger than itself. Found at depths of up to 10,000 feet, the gulper eel uses its bioluminescent tail to lure prey close before snapping them up in a single gulp. Despite its intimidating appearance, the gulper eel is a delicate swimmer, relying on its massive jaws and light-producing lure to survive in the deep sea's vast emptiness.

93. The Baltic Sea Anomaly's True Nature

In 2011, sonar images revealed a strange, disc-shaped structure on the floor of the Baltic Sea. Dubbed the "Baltic Sea Anomaly," it sparked theories ranging from an alien spacecraft to a natural geological formation. Divers who investigated the site reported unusual magnetic interference and a smooth, stone-like surface that defied easy explanation. Scientists later proposed it was likely a glacial deposit from the Ice Age, but the Anomaly remains a popular subject for speculation and exploration, blending science with the allure of the unexplained.

94. The Java Trench: A Portal to the Deepest Earthquakes

The Java Trench, part of the Sunda Trench in the Indian Ocean, is a tectonic hotspot where some of the planet's deepest earthquakes occur. In 2004, this region was the epicenter of a 9.1-magnitude earthquake that triggered a devastating tsunami across the Indian Ocean, killing over 230,000 people. Exploring the trench has revealed the complex dynamics of subduction zones, where one tectonic plate dives beneath another.

95. The Hidden World of Submarine Canyons

Submarine canyons, often larger than their land-based counterparts, carve through the ocean floor, transporting sediment and nutrients to the abyss. One of the most dramatic examples is the Zhemchug Canyon in the Bering Sea, which is deeper than the Grand Canyon. These underwater gorges are teeming with life, from schools of fish to deep-sea corals. Submarine canyons also act as highways for nutrients, supporting ecosystems both at depth and along the continental shelf. Their complexity rivals any terrestrial landscape, yet they remain some of the least explored regions on Earth.

96. The Yonaguni Monument's Geological Debate

The Yonaguni Monument in Japan continues to baffle scientists and archaeologists alike. Some believe this submerged structure, with its sharp angles and stair-like terraces, is evidence of a lost civilization, possibly dating back 10,000 years. Others argue that the monument is a natural formation, created by underwater currents eroding sandstone. Divers exploring the site have reported strange carvings and pathways that hint at human involvement.

97. Deep-Sea Anglerfish: Monsters of the Abyss

The anglerfish, with its grotesque appearance and glowing lure, is one of the most iconic inhabitants of the deep ocean. Female anglerfish can grow up to three feet long, while males are tiny parasites that fuse with the female's body, sharing her bloodstream and becoming little more than sperm providers. Found at depths of 3,000 to 6,000 feet, anglerfish use their bioluminescent lures to attract prey in the pitch-black environment. Their adaptations highlight the extremes of survival in the ocean's darkest corners.

98. Underwater Brine Pools: Lakes Beneath the Sea

In the Gulf of Mexico and the Red Sea, researchers have discovered brine pools—underwater lakes with their own ecosystems. These pools, filled with hyper-saline water, form boundaries so distinct they resemble shorelines. While the surrounding seawater teems with life, the brine itself is deadly to most organisms. However, extremophiles, such as brine shrimp and bacteria, thrive in these pools. Studying brine pools gives scientists a glimpse into life in extreme environments and offers clues about how life might survive in similar conditions on other planets.

99. The Mariana Trench's Challenger Deep

The Challenger Deep, the deepest known point on Earth, lies nearly 36,000 feet below the surface of the Pacific Ocean. Only a handful of humans have descended to its crushing depths, where the pressure is over 1,000 times that at sea level. In 2012, filmmaker James Cameron made a solo descent, discovering strange life forms like amphipods and jellyfish-like creatures. More recent explorations have revealed even more bizarre species, including microbial mats that thrive in methane seeps. The Challenger Deep remains one of the least explored and most enigmatic places on Earth.

FINAL CHAPTER: THE ETERNAL ABYSS

As we close this journey into the ocean's depths, we're reminded that the mysteries of the sea are as endless as the waves themselves. The ocean has revealed much to us—its alien creatures, hidden treasures, and unimaginable landscapes—but for every question answered, countless more remain. The ocean is not merely a body of water; it is a chronicle of Earth's history, a living force shaping our planet's future, and a vast unknown that continues to challenge our understanding of life itself.

From the ancient shipwrecks that whisper stories of the past to the unfathomable trenches that beckon explorers into darkness, the ocean holds a power that is both humbling and awe-inspiring. Its life forms have defied the harshest conditions, thriving in places where survival seems impossible. Its currents connect continents, shaping climates and sustaining ecosystems. And its mysteries—from the strange sounds that echo through its depths to the lost cities swallowed by its tides—remind us of the limits of human knowledge.

But the ocean is not just a frontier of discovery; it is also a warning. As humanity encroaches on its fragile ecosystems with pollution, overfishing, and climate change, the ocean is sounding alarms. The vibrant coral reefs that sustain marine life are bleaching and dying, ancient currents are shifting, and the balance that has endured for millennia is at risk. To truly respect the ocean is to protect it, to recognize that our survival is intimately tied to the health of the seas.

And yet, the ocean's resilience offers hope. It regenerates, reclaims, and renews itself in ways we are only beginning to understand. It challenges us to be stewards of its treasures, to explore it responsibly, and to honor its mysteries without exploitation.

As you leave the pages of this book, remember that the ocean is a mirror of the human spirit: deep, enigmatic, and unrelenting. It calls us to venture further, to ask bold questions, and to embrace the unknown with wonder and respect. The abyss is not just a place; it is a reminder of the boundless curiosity that defines us.

Though our journey into the ocean's depths ends here, the story of the abyss is far from over. For every wave that crashes on a distant shore, for every current that carves a path in the darkness, and for every spark of bioluminescence that lights the deep, the ocean whispers: there is more.

May we always listen, and may we always seek.

www.ingramcontent.com/pod-product-compliance
Lightning Source LLC
Chambersburg PA
CBHW071105240526
45469CB00006BD/2329